The Ant Family-
Fun Facts You Need To Know:
Third Grade Science Series

SPEEDY
PUBLISHING

Speedy Publishing LLC
40 E. Main St. #1156
Newark, DE 19711
www.speedypublishing.com

Like all insects, ants have six legs. Each leg has three joints.

Ants are social insects. There are about 20,000 different species of ants and they can be found almost anywhere.

Ants can carry 20 times their own weight. If a kid was as strong as an ant, she would be able to pick up a car!

Most ants live in colonies which can be extremely large consisting of millions of members.

Queen ants are the head of ant colonies, they lay thousands of eggs. Queen ants can live up to 30 years, the longest of any insect.

Bullet ants, have the most painful sting of any insect. It's sting can last up to 24 hours.

Ants have been able to survive on the earth for more than 100 million years.

Ants communicate using pheromones, sounds, and touch. A pheromone is a secreted or excreted chemical factor that triggers a social response in ants.

16505595R00020